CREATIVE SOLUTIONS PRESS

Test Administration Manual: *Athletic Milieu Direct Questionnaire (AMDQ)*

Screening Test for Female Collegiate Athletes with Eating Disorders/Disordered Eating

David R. Black
Professor Emeritus
Purdue University

Daniel C. Coster
Professor
Utah State University

REVISED FEBRUARY 2018

Copyright © 2018 by Creative Solutions Press, LLC

All rights reserved. No part of this publication may be reproduced, distributed, or transmitted in any form or by any means, including photocopying, recording, or other electronic or mechanical methods, without the prior written permission of the publisher, except in the case of brief quotations embodied in critical reviews and certain other noncommercial uses permitted by copyright law. For permission requests, write to the publisher at the address below.

Creative Solutions Press, LLC
3724 Capilano Dr.
West Lafayette, IN 47906
https://www.creativesolutionspress.com

Disclaimer and Indemnity

The screening test recorded electronically is intended for use only by Creative Solutions Press, LLC. to provide information about potential ED/DE status of the athletes. Failure to explicitly follow the procedures described in the manual for testing may lead to erroneous scores and invalid results and may result in legal exposure. Creative Solutions Press, Inc. hereby disclaims all liability because of carelessness and not adhering to instructions and procedures described in this manual. By participating in this process, you are agreeing to hold harmless Creative Solutions Press, LLC., it agents, officers, directors, and employees from any liability because of any misuse of these procedures described in this manual. You further agree to indemnify Creative Solutions Press, LLC., its agents, officers, directors, and employees from any claims, complaints, and liability resulting from misuse of the manual procedures. You are further advised that Creative Solutions Press, Inc. is a private entity and is NOT associated with or a part of Purdue University or Utah State University.

Printed in the United States of America.
Publisher's Cataloging-in-Publication data
Black, David R. and Coster, Daniel C.
Test Administration Manual: Athletic Milieu Direct Questionnaire (*AMDQ*): For Screening Female Collegiate Athletes with Eating Disorders/Disordered Eating / David R. Black and Daniel C. Coster.

ISBN: 978-1-387-59160-2

First Edition

Tribute to Christine (Tina) M. Bonci, MS, ATC

Christine (Tina) Bonci was a tireless champion and advocate for women's health, well-being, and sports. She dedicated her career to treating her athletes, mentoring her staff, and laying the groundwork for comprehensive care of the physical, psychological, and emotional well-being of those in her care. Her actions and influence have created a culture of change to help athletes, coaches, and staff respect, not neglect, their health and bodies.

One of Tina's major professional distinctions, among many, was being the lead author on National Athletic Trainers' Association Position Statement: Preventing, detecting, and managing disordered eating in athletes that was published in the *Journal of Athletic Training*. Her personage and professional contributions will be missed due to a tragic loss in 2014 at the age of 60 to a rare form of cancer.

Leslie Bonci, MPH, RDN, CSSD, LDN

Dedication

This manual is dedicated to the female collegiate athletes who jeopardize their health and wellbeing to enhance their athletic performance and to healthcare providers interested in changing the athletic milieu, so this is no longer necessary. Detection of these issues brings hope for ameliorating these problems and optimistically, to result in their eradication.

Acknowledgments

Sincere gratitude to the National Athletic Trainers Association Research and Education Foundation for funding research to aid in the development of *AMDQ* Screening Test to detect athletes who suffer from eating disorders or disordered eating.

Appreciation also is expressed to the reviewers who provided invaluable feedback and suggestions regarding the clarity and functionality of the manual for Test Administrators and proctors. The following individuals served as reviewers: Leslie J. Bonci, MPH, RD, CSSD, LDH, School of Dental Medicine at the University of Pittsburgh, Pittsburgh, PA; Toni M. Torres-McGehee PhD, SCAT, ATC, Associate Professor/Graduate AT Director, University of South Carolina, Columbia, SC; and Maya Miyairi, PhD, Assistant Professor, Department of Kinesiology & Health Science, Utah State University, Logan, UT.

Notes

Table of Contents

Those Who Should Read this Manual: 1
Purpose of the AMDQ: 1
Purpose of Manual and Test Standardization: 3
Venues of Administration: 6
Items Included in the Scoring Package 6
Verbatim Administration Instructions: 6
   Group and Individual Administration: 6
      Instructions: 7
   Directions to Test Administrators and Proctors: 12
      Minimizing Distractions: 13
      Supplies Available: 13
      Location of Conveniences: 13
      Assessment Accommodations: 13
      Professional Responsibilities: 14
   Qualifications of Athletic Department Representative and Test Administrator and Proctors: 14
   Before Test Administration Procedures: 15
      Group Size 15
   Student Preparation 16
   During Test Administration Procedures: 16
   After Test Administration Procedures: 17
      After Testing Procedures 17
   Repeat Testing 17
Possible Limitations of the AMDQ 19
Delimitations of the AMDQ 21
References to Review 21
Original Research and PST Test Administration Manual 22
Review Articles (oldest to newest) 22
Conclusions 23
Manual References 24

Notes

AMDQ TEST ADMINISTRATION MANUAL

Those Who Should Read this Manual:

Those who should read the *AMDQ Manual* are those in the Athletic Department responsible for assessments of female collegiate athletes and Test Administrators and proctors, and those on committees deciding on the selection of screening tests to be used with this specific population.

Selection of Test Administrators and proctors is the responsibility of the Athletic Department.

Test administrators and proctors should NOT reveal that this is a screening test for eating disorders/disordered eating among female collegiate athletes.

Purpose of the AMDQ:

- The *Athletic Milieu Direct Questionnaire (AMDQ)* was developed specifically for female collegiate athletes (see section on **References to Review** below). The development of the *AMDQ* took approximately 2 decades. The test was not a revision of a conventional, commercial test for the general population to screen for eating disorders because athletes, performance requirements, and the athletic milieu are unique, and hence, a revision of a screening test for the general population seemed inappropriate.

- The *AMDQ* was named because it was intended to be a "direct" measure of Eating Disorders/Disordered Eating (ED/DE) among female athletes in that questions were developed based on the 3rd and 4th editions of the *Diagnostic Test of Mental Disorders* (*DSM*, *Diagnostic Tests of Mental Disorders*, 1987, 1994, respectively) and taken from other tests used in the assessment process. (See **Possible Limitations of the AMDQ** section, bullet 1 on p. 19). Direct meant the item was to measure the criteria listed in *DSM* for ED, but in the subtlest way without obscuring and thus diminishing the measurement ability of the item (more detailed information about the *DSM* criteria and test development is available in Nagel, Black, Leverenz, and Coster (2000). The *AMDQ* was designed to emphasize prevention. Items were developed and included to identified Disordered Eating (DE). The American College of Sports Medicine in a report authored by Otis, Drinkwater, Johnson, Loucks, and Wilmore (1997) defined DE as "a wide spectrum of harmful and often ineffective eating behaviors used in attempts to lose weight or achieve a lean appearance." The term was operationalized to mean that respondents did not meet the criteria for an ED, but their behaviors were serious and unusual enough that they deserved professional attention, so these behaviors did not become deleterious and lead to an ED. Essentially, those classified as DE met a few ED criteria, but only a couple of criteria under each major category for an ED (see Nagel et al., 2000, p. 434 for full details).

- The *AMDQ* has a distinct advantage in that it can be administered to large groups and requires fewer Test Administrator skills and competencies in comparison to a more "indirect," companion sister screening test. the *Physiologic Screening Test* (*PST*, Black, Larkin, Coster, Leverenz, & Abood, 2003). Because of its more direct nature, sensitivity and specificity may be reduced for the *AMDQ* (but both are within acceptable limits according to epidemiological standards (e.g., Black & Johnson, 2015).

- The *AMDQ* is a "test" designed to be administered in a classroom setting. The *PST* is individually administered and requires sophisticated anthropometric or physiological measurements. These measurements are well within the capabilities and training of a qualified athletic trainer, sports registered dietician, sport psychology consultant, or other members of the sports healthcare team.

- The *AMDQ* is not an objective test used to assess content like a midterm or final exam, but rather focuses on the pressures of the athletic environment ("milieu") and how these pressures and priorities affect dieting and weight loss behaviors. It was designed scientifically as a screening test to quickly, reliably, and validly identify, NOT DIAGNOSE, female university athletes with eating disorders/disordered eating.

- Comparison of the *AMDQ* to other screening tests specifically designed for female college athletes can be found in several published articles cited below that are listed in this manual. We encourage those responsible for assessing athletes (individuals or committees) to read these articles, so they can make an informed decision as to which screening test is best suited for them, their circumstances, and matches the qualification of the person administering the test. Informed decisions are best because they safeguard against "group think" and what is traditional, customary, or the status quo and administering tests not specifically developed for female collegiate athletes.

Purpose of Manual and Test Standardization:

- The *AMDQ* is a "standardized" test and a companion or sister test to the *Physiologic Screening Test (PST*, see pp. 22 and 23). One definition of a standardized test is that it is administered and scored in a consistent manner in a "standard" fashion (Standardized Test, 2018). It is imperative that the questions, conditions for administration, scoring, and interpretation are the same for all respondents. All respondents are entitled to the same opportunities and equality in how the test is administered. Uniform procedures are used to administer the test in the same way to each person who takes the test. There should be no deviations from any of procedures for any reason pertaining to test administration.

- Administering a test in a consistent manner is like conducting a "controlled, scientific experiment," where nothing differs among participants before, during, or after testing no matter when the test is administered, on the same or a different day. Nothing changes; everything is the same. Administering a test consistently maintains or increases reliability and validity of the test, two important psychometric features of tests. Consistency helps also to reduce test bias and to provide accurate and meaningful test results.

- From a tests and measurement perspective, the objective is to find a person's true score and to reduce error or random or systematic effects that would take away from an "accurate" score if the test were administered repeatedly over a short period of time. Within a certain range of variation, the same score would be achieved every time.

- One way to increase chances of a true score is for Test Administrators and proctors is to be familiar with this manual and the way the test is intended to be administered. The manual and procedures for test administration must be studied and practiced in advance of the testing period. The administrators must precisely and carefully read the directions for administration, the test questions, and familiarize themselves with the location of where the test will be administered well in advance of testing. For example, in terms of the test environment, if desks are stationary, seat athletes with a vacant seat between them. The priority should be to use space to foster independent responses. It may be preferable to recruit proctors from your university's testing center to encourage truthful answers. Test Administrators and proctors from the Athletic Department may be intimidating to the athlete or the athlete may be inclined to please them and answer in ways that are socially acceptable or different than how they honestly feel, which is known as "response bias." Response bias diminishes true score, increases error, and reduces the likelihood of accurate and meaningful test results.

- Another way to help in obtaining accurate responses is to eliminate external environmental issues (e.g., distractions; different, vague, or incomplete instructions; seating proximity; noise; temperature, etc.). These external factors may occur before and during a test and can reduce the amount or value of the information that can be gained from the test. The objective is to control these factors, so the assessment is valid and meaningful. To reduce the effects of environmental factors on testing, it is recommended athletes arrive before the start time of the test.

- As a Test Administrator, you are the most critical part of the testing process and in producing accurate results; therefore, there is no place for a casual or nonchalant attitude toward test administration. This type of attitude usually produces meaningless results and results in valuable human and economic resources being wasted. Persons with these attitudes should disqualify themselves immediately as Test Administration or proctor.

- The *AMDQ* is not a test of reading ability. The reading grade level was calculated at 5.27 using the Gunning Fog Formula. Translated, this means the test is suitable for someone who reads at the beginning 5th grade level.

- The *AMDQ* is a screening test only and has NO DIAGNOSTIC CAPIBILITIES. Only a qualified professional specializing in eating disorders/disordered eating can make a diagnosis. More later about who is considered a qualified professional.

- Administration of the *AMDQ* requires that the athlete feels safe and secure with the Test Administrator as well as with the environment in which the test is administered.

- The *AMDQ* can be used to classify an athlete as either meeting or not meeting the criteria operationalized for definitions of 1 of these 3 categories: eating disordered (DE), disordered eating (DE), or no eating disorder/disordered eating (OK). To increase the validity of the test, only ED/DE and OK are provided to the person requesting test results.

Venues of Administration:

- The *AMDQ* is in English only. No translations are available in other languages currently. Translations are underway that soon will be available in Spanish, German, and French.

- The *AMDQ* is intended for group administration, but could, if necessary, be administered individually for make-up purposes or in a counseling setting. Respondents will complete the *AMDQ* online via a web link. Ideally, all respondents will take the test within a group setting at one time and in the same location. Internet access will be needed. A secure (hardwired connection rather than using Wi-Fi) is preferable. Respondents also will need a laptop, desktop, or smartphone at the testing location that can access the *AMDQ* link.

Items Included in the Scoring Package

The following items will be included in your scoring package. (The package may be purchased at https://www.creativesolutionspress.com/products/amdq.)
- A Microsoft Excel file containing your survey ID and participant access tokens.
- A reproducible set of the Administration Instructions.
- A printable DO NOT DISTURB SIGN.
- Website address with the *AMDQ* Test.
- Scoring request form available at htts://bit.ly/cspamdq.

Verbatim Administration Instructions:

Group and Individual Administration:

The following paragraphs are printed and provided to each athlete. A reproducible copy of these instructions will be provided, following purchase of the access codes.

The verbatim instructions are provided below for review and familiarization prior to the test:

AMDQ TEST ADMINISTRATION MANUAL

Instructions:

"Please open a web browser such as Mozilla Firefox or Google Chrome and logon to the test at http://bit.ly/amdqtest. Enter the access token you were provided and then click on Continue.

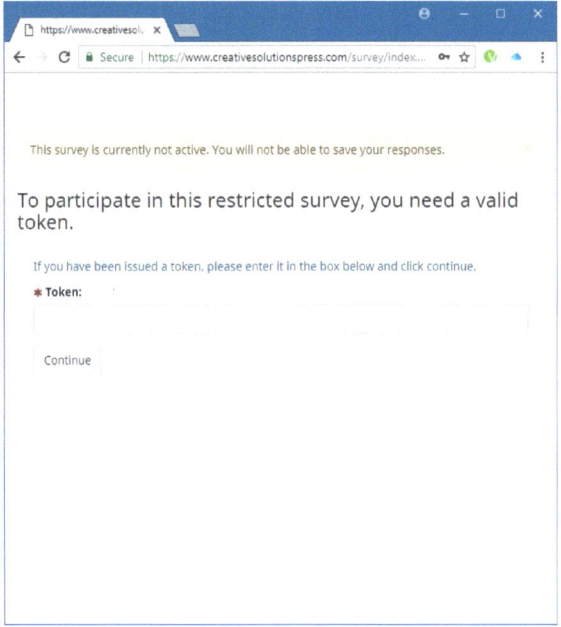

Once the instructions appear, please stop for further instructions."

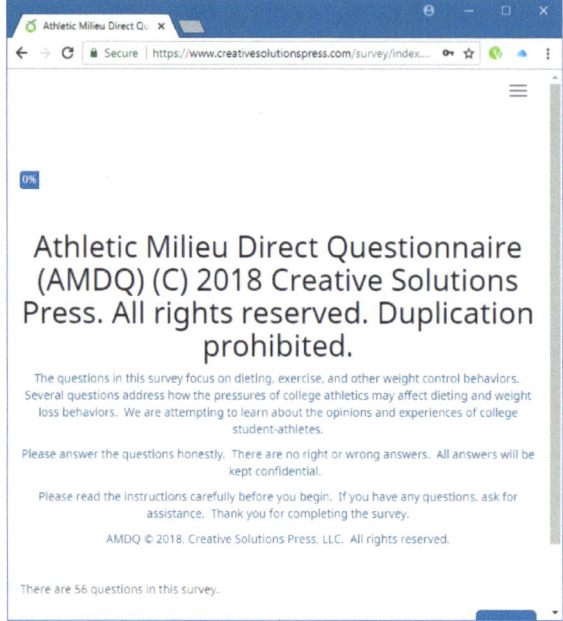

(CHECK TO BE SURE THAT THE ONLY INFORMATION THAT APPEARS ON THEIR SCREENS ARE INSTRUCTIONS.)

Then state:

"Please read the instructions carefully and follow along as I read them aloud. Make certain all personal belongings are out of reach. You may place them at your feet, under your chair, or on the floor in front of the classroom."

OPTIONAL DEPENDING ON ROOM LAYOUT: "Be certain desks are spread out adequately, so you will not be distracted by the pace or responses of others."

"Turn off all electronic devices or silence them. Close all windows on these devices except for your web browser."

AMDQ TEST ADMINISTRATION MANUAL

There are no correct or incorrect answers and the test is not timed. Please answer items honestly and to the best of your knowledge. All your responses are confidential and will be saved on a database.

If you wish to use the restroom, please do so now. We will not begin until everyone is seated.

Each person has been provided a Survey ID and a unique access token for the test. Please retain these numbers in case you are asked to take the test again.

The 56 questions on the survey, which take approximately 20 minutes to complete, focus on dieting, exercise, and other weight control behaviors. Other questions address how the pressures of college athletes may affect dieting and weight loss behaviors. We are attempting to learn about the opinions and experiences of college-student athletes.

At the bottom of each page is a "Next" button. Click on it after you have filled out each of the questions. If the button is not clickable, you have not answered the question on that page. If you need to go back and change a previous answer, click on the "Previous" button.

On the first page of the test, you will be asked to enter the Survey ID _____. Please enter it now. And click on the next

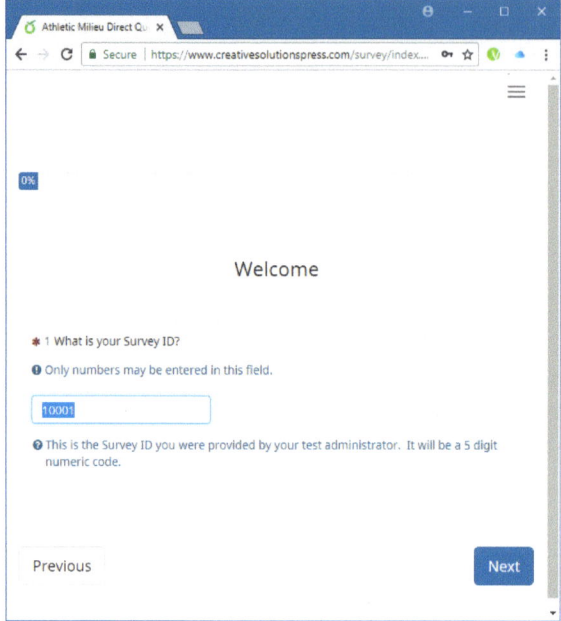

If you receive the message "One or more mandatory questions have not been answered. You cannot proceed until these have been completed.", click on the close button. You will then be able to complete the question and advance.

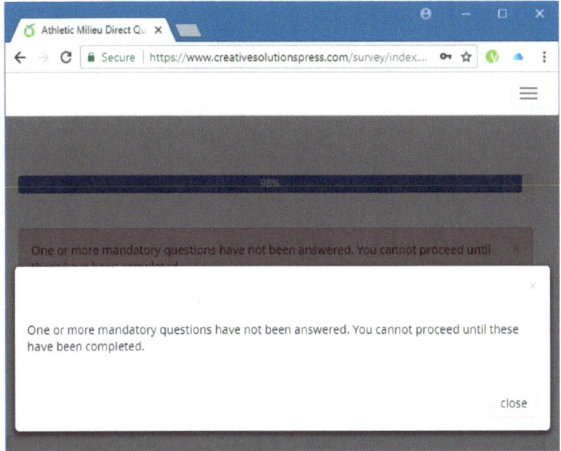

At the bottom of the final page of the test is a "Submit" button. After you have completed every question to your satisfaction, click on the Submit button.

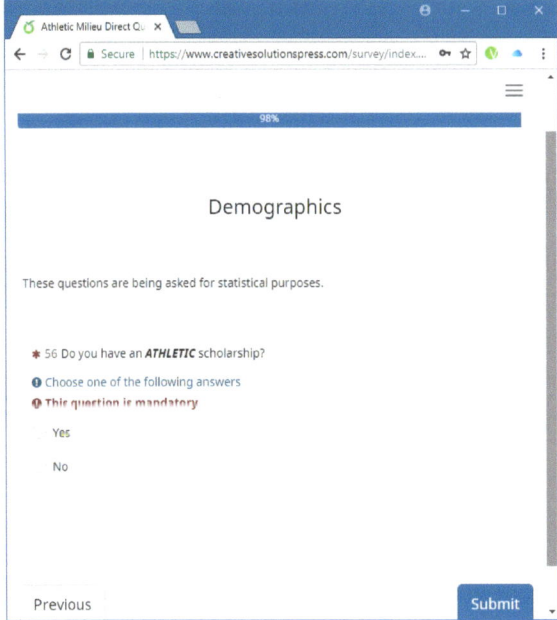

You will receive a message indicating your answers have been recorded. Please signal a proctor that you have completed the test. They will verify that it has been recorded and then you are free to leave.

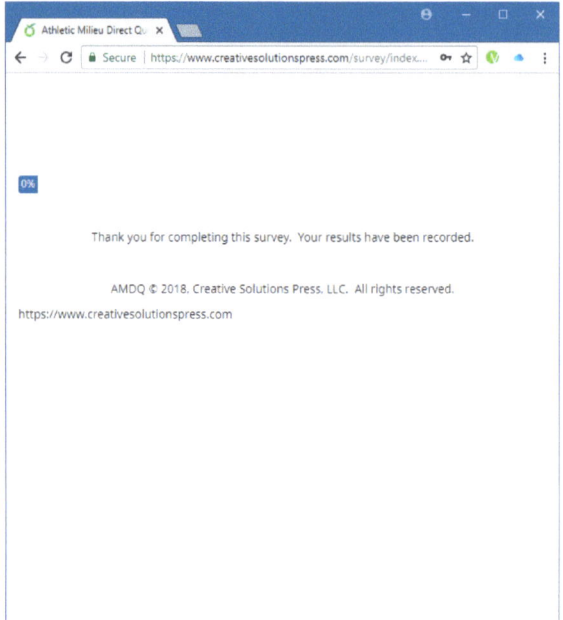

Directions to Test Administrators and Proctors:
Try to limit the number of participant questions asked and answered because the *AMDQ* test questions are written clearly. The focus is on respondent answers. It is imperative to reduce bias by the Test Administrator by not interpreting *AMDQ* questions.

Be discreet while observing test-taking behavior. Be mindful of irregularities or unusual reactions. If someone is insistent on reviewing the answers and rate of completion of their teammate, ask them politely to change seats. If the test questions create anxiety or other observable psychological reactions, speak to the person privately (perhaps, outside the classroom or in the corner of the room where you will not be overheard). Provide reassurance and continue the test in a soothing environment where the test is administered individually. Another option is to reschedule the test at the soonest possible opportunity not allowing an inordinate

AMDQ TEST ADMINISTRATION MANUAL

amount time for reflection about prior responses or about items yet to be unanswered.

Minimizing Distractions:
- Quiet environment
- Clean
- Erased blackboard/whiteboard with no notes. Cover any information on the walls, bulletin boards, and chalkboards/whiteboards that might be distracting
- Safe
- Secluded, if possible, a classroom free from foot traffic outside the classroom
- Comfortable temperature
- No sun glare through windows
- No construction noises
- Quiet ventilation system
- No flickering lights
- Turn all unneeded electronics off (e.g., TVs, projectors, etc.)

Supplies Available:
- Tissue
- Paper towels
- "DO NOT DISTURB" signs to hang on each entry door
- Have enough computers or tablets for each participant to use without there being a necessity to share or wait on others to finish

Location of Conveniences:
- Specify the location of restrooms, drinking fountain, and vending machines. Encourage the use of these conveniences before the test begins or afterwards, not during the test.

Assessment Accommodations:
Testing should be scheduled by the person in athletics in charge of assessments at a time that encourages maximum participation. If the test is not included and time is not allotted during the pre-participation examination, it will be necessary to check athlete schedules and/or inquire about their availability to take the test at

another time and another location. If necessary, because it is impossible to schedule all athletes at the same time, consider scheduling test administration periods sequentially to reduce opportunities to discuss the test between periods. Therefore, it is preferable to test everyone at once as was done during test development. It also is important to select test periods unencumbered by time constraints and other students arriving to use the classroom and walking into the test site.

Professional Responsibilities:

Anyone administering tests to athletes is expected to assume responsibilities presented in the codes and standards of the professional education documented in the *Codes of Professional Responsibilities in Educational Measurement* (National Council on Measurement in Education, 1995). Principles of conduct include, but are not limited to, safety, health, and welfare of respondents; knowing as well as complying with applicable state and federal laws; and performing in ways that are honest and with integrity, due care, and fairness. There also is an obligation to understand the testing procedures prior to administering the test, to administer the test according to the prescribed instructions mentioned in this document, and to avoid anything that would invalidate the test results.

Qualifications of Athletic Department Representative and Test Administrator and Proctors:

The person in Athletics responsible for assessment of athletes has the following responsibilities:

    A. Identify her/himself as person to whom all correspondence should be addressed concerning the *AMDQ*. This would include all contact information: name, address, email address, office and cellphone number, and the best time to be reached.

    B. Order the number of website access codes required for administration of the test for each athlete taking the test.

    C. Provide a request to college/university for test administration assistance far in advance of the test date(s). If no service is available, select the persons (Test Administrator and proctors from the Athletic Department)

least familiar with the group of athletes tested. Students seem to perform better when the test is administered by someone with whom they are unfamiliar.
D. Decide on how many manuals are needed.
E. Provide those administering the test with a copy of the test manual.
F. Interview all those involved in the test administration, so they know their responsibilities and importance of their roles in obtaining meaningful results. It is important to be clear about what is required of them. Answer all questions and refer to the manual when providing answers. Read the part of the manual aloud that pertains to the question or concern, if it is addressed in the manual.
G. Ask for the manuals to be returned at the time of the interview.

Ideally, both the person responsible in the Athletic Department for testing and the Test Administrator and proctors should have a rudimentary knowledge or be trained in test development and psychometrics. The minimum requirement is someone who has experience administering objective tests in a classroom setting and operating computers. Proper selection of the Test Administrator and proctors is imperative to the successful administration of the test and meaningful test results.

Before Test Administration Procedures:
Group Size

Large groups of athletes can be accommodated. The only restriction is the size of the classroom. It is important to select a site where athletes can sit one chair apart (mentioned earlier). The ratio of Test Administrator and proctors to the number of athletes being tested is 1 administrator to every 25 students. Proctors should be added as the groups get larger. For example, between 26 -- 50 students, there should be 1 administrator and 1 proctor; between 51 – 75 students, there should be 1 administrator and 2 proctors; and 76 -- 100, 1 administrator and 3 proctors, etc. It is anticipated that groups will never exceed 150 athletes.

Student Preparation

Test Administrators will provide athletes with rudimentary information: when, where, and time the test will be administered, how the test will be administered (in groups or individually), what the test will assess (see slides), what is expected (reading instructions carefully, answering honestly, reading each item carefully, and not rushing to finish), how will it be scored and by whom (electronically by an independent company), who will see the results (only those authorized by the athletic department on a need to know basis and those who provide direct care to athletes, but never to coaches or those in direct control of or contact with the athletes or anyone in authority who might jeopardize team status).

During Test Administration Procedures:

During testing, participants should log into the survey website. No programs should be open except for the web browser (e.g., Mozilla Firefox, Google Chrome, Microsoft Edge, etc.) and no extra tabs should be opened.

The Test Administrator and proctors should move around the room (not cluster together at the front of the classroom or other locations) to ensure no irregularities and make certain students are following instructions and that orderly progression in answering question is occurring. It is extremely important to identify students who appear to be answering randomly by noting if patterns appear in responses or if they seem rushed to finish. If either behavior is observed, the student should be removed from the classroom with as little disturbance as possible to the group, counseled privately and quietly outside the classroom and asked to retake and complete a new test. A commitment from the student should be sought to take the test fairly and to answer questions honestly. Test Administrator and proctors must be engaged and interested in the respondents and testing procedures.

If the student has trouble understanding what to do especially as it pertains to instructions, assistance should be provided immediately. Again, provide no or minimal assistance regarding responses. Restating what the respondent asks is permissible. Under no circumstances is it permissible for fellow students to aid or to help one another. It is imperative that all athletes answer every question. The software is programmed so this will occur. All students must submit their tests and not logout beforehand.

After Test Administration Procedures:

As soon as students finish, they may be dismissed. They should be reminded to leave quietly and inconspicuously without disturbing others.

After Testing

After all groups have taken the assessment, submit scoring request form at https://bit.ly/cspamdq. Results will be returned in 5 – 7 business days.

Repeat Testing

Once a report is received from Creative Solutions Press, LLC with results for each athlete, the Athletics person has three options. Results are computer scored and divide respondents into 2 groups, ED/DE and OK. Most respondents will be identified as OK and a (disproportionately) smaller group will be identified as ED/DE. This is a common result of screening tests and what to expect.

Option 1: Based on the *AMDQ* results, those who are identified as ED/DE could be immediately referred to a healthcare provider with expertise in ED/DE among athletes. This expert would diagnose whether the athlete was ED/DE or OK and decide on the treatment course for those diagnosed as ED/DE. This course of action potentially could be overwhelming and relatively expensive based on the number of athletes screened and the number of qualified service providers available. There is no suggestion that this is an inappropriate option. It depends on the resources available. At the very least, a series of presentations by experts in sports psychology and sports dietetic could be helpful and useful rather than enrolling all those diagnosed in treatment. This approach could be considered as a supplement to treatment.

All screening tests, unless they are 100% accurate (a rarity for most tests used routinely in healthcare) divide all respondents into those who are identified by the test with the problem and those without it. Some (about 23% based on *AMDQ* test development results) will be false positives (those who test positive but are not ED/DE) and some (about 20% based again on test development results) will be false negatives (test negative but are ED/DE). Usually, small numbers of respondents fall into these two categories, but not insignificant numbers. Simply put, the *AMDQ*, like any quality screening test, will correctly classify the largest percentage (~78% overall for the *AMDQ*) as ED/DE or OK, but regrettably, some (~22%) will be misclassified. This is the reason for repeat screening. The *AMDQ* correctly classified approximately 4 out of 5 athletes who were truly exhibiting ED/DE and about the same fraction who were OK. In epidemiologic terms, the sensitivity for the *AMDQ* was 80% and the specificity was 77%., two very important terms for evaluating the utility of a screening test. See Black and Johnson (2015) for further information about screening tests.

Option 2: A second option is rescreening with the same test or administering the *AMDQ* again. This option can be more efficient and cost-effective than *Option 1*. The key is to let enough time elapse after the initial testing; otherwise, the results will vary little. When screening a second time with the same test, the focus is on those who tested negative the first time according to the test, which includes false negatives and those who are OK. Retesting those in these 2 categories produces the largest groups to retest because those who are OK is the largest category of all those tested. The second test will identify additional athletes who on this second occasion will test as true positives and increase certainty about those who are OK. The result is additional athletes who tested positive and they would be referred to a professional for diagnosis and treatment.

Option 3: The third option is to use a completely different test. One option is the *Physiological Screening Test (PST)*, also developed specifically for athletes. This test is available through Amazon.com at http://amzn.to/2EAHqLh. When testing a second time with a completely different test, instead of screening again those the test identified as negative for ED/DE (i.e., false negative and true negatives), the focus is on those who tested positive. The focus this time is on true and false positives to confirm those who are ED/DE and to identity others among the false positives who are ED/DE. There would be greater confidence when using a different test in the number who were true positives and in those who should be seen for diagnosis and treatment.

Selecting any test for screening either initially or a second time can be challenging. A list of references below is provided to help with test selection and to explain more about screening. It is imperative to remember that a screening test is not a DIAGNOSTIC test and it does not provide a diagnosis. It is an indicator of a potential problem. A skilled professional with training in diagnostics would make the decision if an athlete is ED/DE or OK and decide on subsequent steps for intervention

Possible Limitations of the *AMDQ*

The limitations listed are present in any screening test:

- The value of the *AMDQ* is not negated because it was developed based on criteria from *DSM-IIIR* and *DSM-IV*. The diagnostic criteria for anorexia nervosa and bulimia nervosa in *DSM-5* remain largely unchanged from *DSM-IV* (Grohol, 2018, February 21). The major change to anorexia nervosa is that amenorrhea was removed and Criterion A, which now focuses on behaviors like "restricting calorie intake" and the word "refusal" was removed because it implied intention. Other word changes also have been made in Criteria A and B. The only change pertaining to bulimia nervosa was to reduce the frequency of binge eating and compensatory behaviors from a minimum threshold of once a week to twice a week.

- There are several reasons why *DSM-5* was unavailable at the time of the study. The study took several years to complete due to the length of time required to conduct preliminary psychometric assessments related to reliability and validity. The structured diagnostic interview required many hours of practice for interviewers to be certified as proficient coupled with the time required to schedule and interview athletes. The uniqueness of the analyses was challenging because there was no prototype to follow. Multiple considerations were considered in deciding on the best statistical and epidemiologic approach. There were often constraints in recruiting athletes and in scheduling testing because of busy athletic schedules.

The *AMDQ* is still relevant, even though changes in the *DSM* criteria have changed. It should be endorsed and used because it is a screening test or general indicator of ED/DE. The items in the screening test were written were taken from many sources and not specifically tied to the *DSM* criteria.

The items selected were those that statistically discriminated best between ED/DE and OK for the "gold standard," the standardized diagnostic interview. Diagnoses are the reasonability of qualified and skilled professionals. Anything missed by or was too stringent in screening would be further evaluated and weighted in the diagnostic interview with a professional. Retesting options for repeat screening mentioned above would still apply and be useful, but not in making a finer discrimination other than the broad indicators of ED/DE or OK.

- The *AMDQ* is comprised of subjective items. Accordingly, internal and external validity and response bias may be compromised because respondents may not be completely truthful, independent of their ability to detect the purpose of the test. This criticism may have some merit but does not seem plausible because these issues were extensively addressed during test development. Response bias was evaluated on two different occasions. The initial test was evaluated by 20 female senior high school students to avoid a reduction in the collegiate athlete respondent pool. These students were asked to evaluate each item as to whether it would result in an honest answer or an answer that would conform to socially acceptable norms. Based on the feedback received, some items were revised, and others eliminated. This process was repeated with a second group of 39 female high school senior students not a part of the first testing for response bias. Again, items were revised or eliminated. According to final analyses, response bias was minimalized or no longer an issue among the final items used in the test.

- Generalizability/external validity may be an issue since only Division I athletes from one major university were included. However, similar prevalence rates were found with the club athletes and dancers when evaluated with the *AMDQ*, suggesting that at least among these respondents the test was effective for club and elite athletes.

- Content validity might be challenged. However, the final version of the test was evaluated for content validity by soliciting feedback from 3 content experts in the areas of eating disorders and athletes and who were familiar with athletes and athletic domain. Feedback was solicited by using a structured evaluation form. Based on the experts' feedback, modifications again were made to the test. The result was that experts were in 100% agreement about the appropriateness of the items after items were modified or eliminated.

Delimitations of the *AMDQ*

- The *AMDQ* provides a broader spectrum of classification and early intervention or prevention of eating disorders than other commercial tests of eating disorders except for the *PST*. The authors consider this an advantage, but it may limit comparison of results with other tests developed differently.
- The study was limited to female college athletes between 18 – 25 years old. This age group was selected because these age groups are the most commonly represented among college female athletes.

References to Review

As a member of the Athletic Department and a person responsible for test selection and administration or members of a committee with this charge or a Test Administrator, it is important to be informed about the test you are administering and how it was developed. You may have to defend its use and selection to those in the athletic community and to the professionals to whom you may make a referral. Documentation and justification seems to be a requirement, and selection and use cannot be based on custom or recommendations from others in the field. A rationale and informed decision should be the basis of choice and selection.

There are several open access documents that are essential to review. The list below is not exhaustive. All the documents listed are open access. Open access means that the document is free of charge and can be download at the URL address below the citation. The articles have been organized by category. The first 2 articles are the original research conducted to develop and validate the *AMDQ* and the *PST*. They are numbered 1 and 2, respectively and 3 is the test administration manual for the *PST*. The other 3 articles are reviews of the *AMDQ* (and the *PST*) published in the research literature. Alpha characters (A-C) have been used to distinguish them. Both the *AMDQ* and *PST* have received positive reviews.

Original Research and *PST* Test Administration Manual

1. Nagel, D.L., Black, D.R., Leverenz, L.J., & Coster, D.C. (2000). Evaluation of a screening test for female college athletes with eating disorders and disordered eating. *Journal of Athletic Training, 35,* 431-440. http://bit.ly/2soeKjx

2. Black, D.R., Larkin, L.J., Coster, D.C., Leverenz, L., & Abood, D.A. (2003). Physiologic Screening Test for eating disorders/disordered eating for female athletes. *Journal of Athletic Training, 38,* 286-297. http://bit.ly/2CbkIEk

3. Black, D.R., Leverenz, L.J., Coster, D.C., Larkin, L.J., & Clark, R.A. (2010). *Physiological Screening Test (PST) manual for eating disorders/ disordered eating among female collegiate athletes.* Monterey, CA: Healthy Learning. ISBN: 978-1-60679-069-4 http://bit.ly/2Ej1DpB

Review Articles (oldest to newest)

A. Bonci, C.M., Bonci, L.J., Granger, L.R., Johnson, C.L., Malina, R.M., Milne, L.W., & Vanderbunt, E.M. (2008). National Athletic Trainers' Association Position Statement: Preventing, detecting, and managing disordered eating in athletes. *Journal of Athletic Training, 43,* 80–108. DOI: https://doi.org/10.4085/1062-6050-43.1.80
http://bit.ly/2nRVQgn

B. Knapp, J., Aerni, G., & Anderson, J. (2014). Eating disorders in female athletes: Use of screening of screening tools. *Current Sports Medicine Reports, 13,* 214-218. DOI: https://doi.org/10.1249/jsr.0000000000000074
http://bit.ly/2nRtwuL

C. Wagner, A.S., Erickson, C.D., Tierney, D.K., Houston, M.N., & Bacon, C.E.W. (2016). The diagnostic accuracy of screening tools to detect eating disorders in female athletes. *Journal of Sports Rehabilitation, 25,* 395-398. DOI: https://doi.org/10.1123/jsr.2014-0337
http://bit.ly/2Bmfmcs

Conclusions

The purpose of this manual is to reduce test bias and to obtain accurate scores. The test administration practices discussed in this document are based on the accepted codes and standards of leading educational organizations, the latest findings of education researchers, and 4 decades of test administration experience by the authors. By following these practices closely and completely, the Athletic Department and the Test Administrator help athletes to provide the most accurate responses possible under standardized conditions. Therefore, it is incumbent on the Athletic Department to select the most qualified Test Administrators. Test Administrator must prepare themselves to administer the test according to the instructions provided and not to take test administration for granted or to consider it proforma. Prudent selection of Test Administrators and proctors will increase the chances that athletes will get the help they need.

It must be remembered that a screening test is NOT a DIAGNOSIS, but an indicator of a potential risk for ED/DE. Typically, diagnoses are made by qualified professionals in psychiatry or psychology. Athletes who test as ED/DE should be referred to professional who hold these licenses and who are experts in eating disorders and familiar with the athletic milieu.

Manual References

Black, D.R., & Johnson, M.A. (2015). *Handbook for foundations of epidemiology* (3rd ed.). West Lafayette, IN: Creative Solutions Press, LLC. ISBN: 978-1-329-46928-0

Diagnostic and Statistical Manual of Mental Disorders (3rd ed.). (1987). Washington, DC: American Psychiatric Association. ISBN: 978-0-521-31528-9

Diagnostic and Statistical Manual of Mental Disorders (4th ed.). (1994). Washington, DC: American Psychiatric Association. ISBN: 978-0-890-42061-4

Grohol, J.M. (2018, February 21). DSM-5 changes: Feeding & eating disorders. Retrieved from http://bit.ly/2Fn9P5B

Otis, C.I., Drinkwater, B., Johnson, M., Loucks, A., & Wilmore, J. (1997). American College of Sports Medicine position stand: The female athlete triad. *Medicine and Science in Sports and Exercise, 29,* i-ix. doi: https://doi.org/10.1097/00005768-199705000-00037

Standardized test. (2018, January 30). Retrieved January 31, 2018, from Wikipedia: http://bit.ly/2EibnQT

Notes

About the Authors

David R. Black, PhD, MPH, HPPE, CHES, CPPE, CPPT/C, FASHA, FSBM, FAAHB, FAAHE, is Professor Emeritus and a licensed healthcare provider in psychology with expertise in eating disorders and disordered eating in the athletic milieu and a behavioral epidemiologist in the Department of Health & Kinesiology at Purdue University. He has taught courses pertaining to eating disorders/disordered eating, and weight reduction pertaining to athletes and epidemiology for more than 2 decades. One of the components of the courses he taught at both the undergraduate and graduate level focuses on the design, development, strengths, and limitations of screening tests. He also teaches course related to the design, implementation, and evaluation of prevention and intervention programs. He specializes in programs that utilize trained peer helpers. Dr. Black also has the distinction of holding courtesy appointments in Health Sciences, Nutrition Sciences, and Nursing, and the Purdue Polytechnic Institute – Computer Information and Technology.

Daniel C. Coster, PhD, is a research and teaching statistician in the Department of Mathematics and Statistics at Utah State University. Having earned his undergraduate degree with first-class honors in mathematics and master's degree in mathematics from Cambridge University, England, he received his PhD from the Department of Statistics, University of California, Berkeley, in 1986. He joined the faculty at Utah State in 1990, where he became a full professor in 2009. His primary responsibilities at Utah State include teaching graduate and undergraduate statistics courses, both theoretical and applied, and conducting research on the theory of optimal experimental design. Dr. Coster is renowned for his expertise in the design of sample surveys and experiments for health-promotion studies and the analyses of data from such research endeavors.

www.ingramcontent.com/pod-product-compliance
Lightning Source LLC
Chambersburg PA
CBHW041117180526
45172CB00001B/301